Insight Outlook

Insight Outlook

Dr. Albert Hofmann

HUMANICS NEW AGE
Atlanta, Georgia

HUMANICS NEW AGE
P.O. BOX 7447
Atlanta, Georgia 30309
Humanics New Age is an imprint of Humanics Publishing Group.

Copyright © 1986 by Albert Hofmann
 © 1986 by Sphinx Medien Verlag, Basel, Switzerland
 © 1989 (English) by Humanics Limited
All rights reserved.
No part of this book may be reproduced by any means,
nor transmitted, nor translated into a machine language,
without written permission from Humanics Limited.

First Edition

PRINTED IN THE UNITED STATES OF AMERICA

Library of Congress Cataloging-in-Publication Data

Hofmann, Albert, 1906-
 [Einsichten, Ausblicke. English]
 Insight, outlook / Albert Hofmann.
   p.   cm.
 Translation of: Einsichten, Ausblicke.
 ISBN 0-89334-116-9
  1. Religion and science—1946- 2. Mysticism. 3. Nature-
 -Religious aspects. I. Title.
 BL262.H6413 1988
 215—dc19   88-39854
                                                          CIP

Photographs by RANJIT AHUJA
Cover and Text layout by PURNA AHUJA

Contents

Editor's Foreword	vii
Introduction	xiii
The Interrelation Between Inner and Outer Space	1
Security in the Natural Scientific-Philosophical View of Life	31
On Possession	59
Botanical Reflections on the Death of the Rain Forests	71
The Sun, a Nuclear Power Station	79

Editor's Foreword

The mark of true great thinkers has always been their scholarship and academic accomplishments, as well as the ability to translate difficult concepts into an accessible form. Dr. Albert Hofmann, Member of the Nobel Prize Committee, Fellow of the World Academy of Sciences, Member of the International Society of Plant Research and the American Society of Pharmacognosy, is such a man.

The book you hold in your hands is the product of this great thinker, possessor of a brilliant mind who happens to think in another language. Not merely a language of phonetics and sentence structure, but a language of thought, of novel inductive and deductive reasoning. As the reader progresses through INSIGHT\OUTLOOK, it becomes readily apparent that the flow of ideas and associations is quite different from what one may have come to expect from routine academic study or scholarly musings.

Dr. Hofmann draws from his immense knowledge of the physical and social sciences to weave a picture of reality, both objective and spiritual, that is accessible to all. Touching first on one realm of conjecture and then another, he draws us into a kinetic but very controlled dance of reasoning. This wonderful train of thought spirals around and around, always relevant and revealing, until finally the dance ends and the entire concept emerges complete and multi-dimensional. As always, Dr. Hofmann not only leads us to places in the mind where we have never been before, but he also bids us to use our new-found insight as a launching area for further intellectual meditation.

Two fundamental insights permeate the observations contained in INSIGHT\OUTLOOK. The first is based on the recognition that the natural scientific concept of life and the world view unveiled by visionary experience are not contradictory as we have been taught to believe, but are indeed complimentary. The second is a series of reflections on the interrelationships of the outer, objective, material world with the inner, subjective, spiritual world. Both appear to be mutually exclusive, yet both are fundamental aspects for the construction of what each individual human being calls reality.

In the first chapter "The Interrelation of Inner and Outer Space", the metaphor of a television broadcast is used to illustrate the production of reality. The outer material world functions as a transmitter, the inner spiritual and creative center of the individual human being functions as the receiver. There is but one transmitter, the outer world, but there are as many receivers as there are human beings. This in turn creates individual, different, and viable realities for every person on earth.

The second chapter, "Security in the Natural-Scientific-Philosophical View of Life", shows how the increasing knowledge of the physical world gained through the explorations of the natural sciences closely parallels our growing awareness of our place on earth, both in the past and in the future. These world-wide insights, which closely resemble visionary experience, can provide us as a species with a better understanding of our place in the cosmos and our duties for the safe-keeping of our planet.

The existential relationship between objects and owners is the topic of the third chapter, "On Possession". Possession becomes property by its judicial recognition and protection. The means by which an object is enjoyed however, involves a subjective appreciation of it, an emotional bond. Objects in this use of the term are understood to include spiritual experiences and relationships. The gift placed in every human being's cradle is the ability to become not only possessor of their immediate physical surroundings, but of their own individual, subjective world.

"Botanical Reflections on the Death of the Forests" reconstructs the processes by which the forests of the world are slowly dying at the hands of man's industrial waste products. The basic concepts of the assimilation of these poisons are available in every elementary textbook on botany, yet having no immediate practical use they are ignored. In this chapter, we are shown how the signs that herald the death of the forests forecast the endangerment of all life on our planet.

The human spirit, our consciousness, is shown to be the highest, most energetic transformation of light in the final chapter "The Sun, a Nuclear Power Plant". The dangers of conventional man-made nuclear power are contrasted with the clean, inexhaustible power of the sun. Throughout our planets evolution, the sun has been the engine that has driven the production of coal, gas, oil, and the hydraulic power of rain, lakes and rivers, even the cleansing and regeneration of the very air we breathe. To capture but a small fraction of this immense, extraterrestrial nuclear power source is shown to be a solution for all our energy problems to come.

INSIGHT\OUTLOOK uniquely opens the readers mind not only to the building blocks of the past, but to the possibilities of the future as well. It has been a labor of love to retain the flavour, the intensity and structure of Dr. Hofmann's work. To this end, special thanks are extended to Dieter Hagenbach for preliminary translations, Jennifer Wilson for consummate production, and Gary B. Wilson for guidance and direction.

Dr. Hofmann makes several referrals to the word "utopia" in the following work. The translation of this word from the Greek is "nowhere". INSIGHT\OUTLOOK is a roadmap to the discovery of a world extant not only in the hopes and dreams of humankind, but also as a guide to a physical reality readily possible in the critical years to come. Dr. Albert Hofmann is our Beatrice and we are the travellers, our hearts and minds the hands and tools required to drag the dream of utopia into existence.

Robert Grayson Hall
Editor

INTRODUCTION

The world is a rotating sphere moving through space while circling the sun. Every one knew this, but then it could be seen when space research programs supplied photographic images a few years ago: the planet earth, a blue sphere, floating free in space.

Since then, I have been fond of evoking this picture in my inner eye before going to sleep. I imagine how I, lying here in my bed, am traveling along there on the surface of the sphere, where so much has happened since it has been quietly moving along the trajectory determined for it in primordial time.

Only after billions of revolutions around the sun, after planet earth had greened and after many more millions of years, when animal life had developed, did the being appear that experiences the world and itself consciously. As one of these beings gifted with consciousness I now observe the blue sphere, on which the drama of mankind takes place, from space through the eye of a camera. The fate of nations, the personal histories that have been acted out there, now separated from today's viewer by the curtain of time! But the images live on in the timelessness in which we all participate through our consciousness: fabulous cultures that flowered in China centuries ago, the world of Greek and Roman antiquity, Alexander the Great's Persian War, the Aztec empire, the Crusades, the Gothic and Renaissance periods, two World Wars . . .

Nothing of this shifting scenery could be seen on the surface of the earth from the cosmic perspective, and neither could be seen the generations of individual humans appearing and disappearing in it. The picture that presents itself from space today has always been the same—the blue sphere, illuminated by the sun, floating quietly through space, unperturbed by time and the fate of mankind.

While this picture is etched on my inner eye with the clarity of a photographic image, I know that I am presently on the dark side of the surface of that sphere, that I am here in my house which sits in a field in the Swiss Jura, in the bedroom, with the window open, breathing the fresh night air mixed with the scent of hay. On the sphere my individual existence vanishes among the billions of humans inhabiting its surface at the moment for a cosmic second. In contrast, here I am, the center of the world, of my world, extending from my room across the countries of the earth to the moon, to the sun, into the infinity of space aglitter with stars.

What then is true, what is real, am I here or there? Is it even possible to ask this question, the answer to which seems so obvious? I think so, for basically nothing is obvious. That so much, almost everything, appears to be obvious to us to is one of the most monumental errors in our mental attitude. Obviousness could be the ruin of the world.

The answer to the above question, I am here in my room and there on that blue sphere is not obvious. It represents a higher truth only understood by someone who realizes that the world on which he stands is a sphere. The only thing true and real to a primitive man is what he can experience directly with his senses, in this case that he is here, on the world, that it is flat with the firmament stretched above it. He only knows a part of the truth.

In the following work I would like to show what this example of nocturnal meditation demonstrates: that, depending on the standpoint of the observer, reality offers quite different perspectives that are not necessarily mutually exclusive. Rather, they add up to an encompassing truth. They contain insights into the essence of our day-to-day reality that I have acquired through experiences in my own life. Thus, they are very personal observations on a central problem of reality that inevitably lead us into the realm of religion.

Everyone is, in fact, his own philosopher, for every human experiences the world in a unique manner according to his own uniqueness and, correspondingly, devises his own personal image of it. Everyone has to make do in his special reality.

The questions asked by children show that we are all already born as philosophers: "Dad, where does the world stop?—When did God make the world?—Why does everyone have to die?"—and the like. They are questions to which we still cannot find the answers in all of the many philosophical treatise, even though they are basic questions relating to our existence.

I still clearly remember a childlike philosophical discussion from my youth that I had with a friend of mine when I was about ten years old. We were on our way to school and were just strolling toward the old city gate when my companion asked me, "Do you still believe in God? Since I noticed that they cheated me with the Christ child (translator's note: in the tradition of German speaking countries, the Christ child plays the role of Santa Claus as the bearer of gifts on Christmas Eve) and that St. Nicholas was none other than my Uncle Fritz, I no longer believe that he exists." I answered that it had to be different with God than with the Christ child and St. Nicholas. The world and people, that only God could have made, really did exist.

That was my proof of God, and it still is today.

Why do children ask such profound questions?—Because the creation, disclosing itself to their fresh senses directly and untouched, does not appear to be obvious to them yet. It is the adults who see it that way through a perception dulled by habit. But that is not how it is, the children are right. They are still living in Paradise because they are still living in the truth. They still perceive the world as it really is, as wondrous.

Adults now only experience amazement about the newest inventions and products of science and technology, about computer-guided rockets, laser discs, space travel, etc. We have every reason to admire these magnificent accomplishments of human ingeniousness, even if some of them do scare us. The tragedy is that we fail to see the secondary, transient character of all of man's works, that we are not aware that science and technology are based on things pre-existing in nature. The chemist works with the matter of which the earth is made; the physicist and the biologist investigate the forces and laws of a transcendental origin that preserve the anorganic universe and animate the world of plants and animals; the technician makes those laws subservient and exploitable.

The origin of the primary world, of the creation with its laws governing the course of the stars and the growth of a blade of grass that existed before man appeared, is beyond rational explanation. The insights of the natural sciences are descriptions of preexisting conditions, they are not explanations. The botanist can describe the shape and color of a flower to the last detail and compare it with other blossoms; the cell physiologist can investigate and vividly depict the mechanisms of fertilization, cell division, and the manner in which this blossom forms organs. But why a flower is the way it is, where its design plan and the laws according to which this plan are executed came from, remains a mystery. The child sees the flower as it is in its totality and thus sees what is essential, namely the miracle. In comparison to this, whatever scientific research might add is of little import.

But it is by no means meaningless. I became a chemist and then studied the chemistry of plants precisely because I was attracted by the mystery of the matter and the miracle of the plant world. The insights into the construction of matter and the chemical structure of the pigments of blossoms and other plants that I acquired in my work did not diminish my amazement about nature, about its workings, about its forces and laws—they increased it. An insight into the internal structure and life processes of natural objects is added to the perception of form and color imparted by the observation of their surface. This results in a more complete picture of their reality, a broader truth.

It could well be that the value and importance of the natural sciences is not primarily that they gave us modern technology, comfort, and material affluence; their actual, evolutionary meaning could be the expansion of human awareness of the miracle of creation. The recognition of the creation as a firsthand revelation, as "the book written by the finger of God," could become the basis of a new, world encompassing spirituality.

Natural scientific research has let us perceive how man is embedded in the entirety of nature and how he represents an inseparable part of it. This knowledge is in conformity with the emotional experience of the mystic and the unity of all living things. It appears that this fundamental truth is now entering increasingly into the general consciousness from these two sides in a complimentary manner.

This opens up a promising view into the future, for the main problems of the present derive from a dualistic awareness of reality. The perception of our natural environment as something separated from man, as an object that can be used and exploited without limitation, has led to an ecological crisis. The newly awakening religious awareness of the unity of man with nature, and only it, could induce the unavoidable.

A personal, childlike perception of nature, to be equated with the mystical experience, as the one source and natural scientific insights as the other, form the basis of the following work. These two complimentary views and insights into the unity of the outer material and inner spiritual worlds, the natural and humanistic sciences, define my philosophy of life. It does not contain any new philosophical insights; rather it is the result of a timely, personal experiencing of old truths. I found security, trust, and safety in it because its main features coincide with the concepts of the great philosophers and their common religious origin.

Rittimatte, Burg i.L.
Switzerland

Insight Outlook

1

The Interrelation between Inner and Outer Space

> Reality is just as magical as the magical is real.
>
> Ernst Juenger
> "Sicilian Letters to the Man on the Moon"

THERE are experiences that most people avoid talking about because they do not fit into the reality of everyday life and defy rational explanation. I am not referring to specific events in the outer world, but to processes within us that are generally dismissed as mere imagination and expelled from memory. In the experience I am talking about, the familiar view of our environment suddenly undergoes a strange, exhilarating, or frightening metamorphosis, appears in a different light, takes on a special meaning. Such an experience can waft over us like a breath of air, or it can affect us profoundly.

I have a very vivid memory of such an enchantment from my youth. I do not remember the year, but I can still recall the precise location where it took place, somewhere along the path through the woods on the Martinsberg above Baden, Switzerland. As I was strolling through freshly greening woods illuminated by the morning sun and filled with bird-song, everything suddenly appeared in an unusually clear light. Had I never looked at it closely enough before, and was I all of a sudden seeing the spring forest as it truly was? It radiated in a glow of eloquent beauty, touching the heart in a singular manner, as though it wanted to include me in its glory. I was filled with an indescribably blissful feeling of belonging and a blessed security.

I do not know how long I stood there enthralled, but I remember the thoughts that occupied me as my radiant condition slowly dwindled and I walked on. Why did this blissful view not linger on having, after all, revealed a reality rendered convincingly by this direct and deep experience? And how could I, urged on by my overflowing joy, tell anyone about my experience, sensing immediately, as I did that I could find no words for what I had seen? It seemed strange that I, a child, had seen something so wonderful, something that adults obviously did not notice. For I had never heard them talk about it—or was this one of their secrets?

During my later youth I had several more of these exhilarating experiences while walking through fields and forests. They came to determine the basic concepts of my view of life by assuring me of the existence of an unfathomable, vivid reality hidden from the view of everyday life.

I included the above description of my childhood visionary experiences in the introduction to my professional autobiography, LSD—MY PROBLEM CHILD, (Stuttgart, 1979), for that mystical experience of reality was one of the reasons I chose to become a chemist. It awakened in me a longing for a deeper understanding of the make-up and essence of the material world. During the course of my professional activities, I encountered psychoactive plant substances that are, under certain conditions, capable of triggering visionary states similar to the spontaneous experiences I have described above. My experiments with mind-altering substances, of which LSD became known throughout the world, led me to the problem of the relationship between substances and consciousness, between the outer material and the inner spiritual world.

What we call reality is obviously an interrelation between inner and outer space. Reality is inconceivable without a subject being, a self to experience it. It is the product of interrelation between a transmitter in the outer space and a receiver in the inner space.

I use the expression outer space in its general, everyday meaning. I am not referring to curved space or the four-dimensional space of theoretical physics. I am talking about three-dimensional Euclidean space. I am talking about nothing other than the empty space that can be filled with material objects.

The inner space is consciousness. Consciousness defies definition, for it is what I need to contemplate what consciousness is. It can only be circumscribed as the receptive and creative center of the spiritual self.

There are two basic, characteristic qualities that define the differences between outer space and inner space. The first: while there is only one outer space, there are as many inner spaces as there are humans. The second characteristic difference is this: inner space is a purely subjective mental experience, while outer space exists objectively.

And now for the definition of reality that I have alluded to here. The reality I am talking about in this context is not a transcendental reality, a reality of theoretical physics capable of being illustrated and explained with the help of mathematical formulas alone. The reality I want to talk about is the reality we mean when we use the expression in everyday life, the world as we humans perceive it with our senses. Reality thus defined is inconceivable without an experiencing being, without an ego. It is the result of an interrelation between material and energetic signals that emanate from the outer world, from the outer space and a consciously experiencing self in the inner space of a human.

To illustrate this, we can compare the process by which reality is created with the generation of sounds and pictures during a television broadcast. The material world in the outer space functions as the transmitter, giving off optical and acoustic waves and supplying tactile, gustatory, and olfactory signals. The receiver is embodied by the deep self of an ego, where the stimulae received by the antenna of the sensory organs are transformed into an image of the outer world in the inner space where they can be experienced mentally.

If either the receiver or the transmitter is missing, no human reality can be created, just as the screen of a television would remain empty in the absence of pictures and sound.

In the following I will try to explain what we know about the physiology of man with regard to his function as a receiver, as well as about the mechanisms of receiving and experiencing reality on the basis of scientific insights.

The antennae of the human receiver are comprised of our five sensory organs. The antenna for optical images from the outer world, (the eye), is capable of receiving electromagnetic waves, and of projecting a picture onto the retina that coincides with the object from which these waves emanate. It is important to realize that the human eye can only receive a very small section from the very broad spectrum of electromagnetic waves emanating from the outer world to depict objects from the outer space. The immeasurable spectrum of electromagnetic waves traveling through the universe extends from wave lengths of one billionth of a millimeter, corresponding to the range of X-rays and ultrashort Y-rays, up to radio waves that are many meters long. Our eyes only respond to a very small range between 0.4 to 0.7 thousandth of a millimeter (=0.4 to 0.7 millimicrons). Our eyes can only receive and perceive this very limited range as light. All the other rays in the boundless domains of electromagnetic waves filling the universe do not exist in human reality.

Within the small spectrum of visible waves between 0.4 and 0.7 millimicrons, our eyes and the receiver in the inner space are capable of differentiating between different wave lengths and recognizing them as different colors.

In connection with our reflections it is important to note that colors do not exist in the outer space. Generally speaking, we are unaware of this basic fact, even though it can be looked up in every textbook on physiology. Objectively, all that actually exists in the outer space is matter transmitting electromagnetic oscillations of varying wave lengths. If an object reflects or transmits electromagnetic waves with a length of 0.4 millimicrons from the light shining on it, we say that it is blue; if it transmits waves with a length of 0.7 millimicrons, we describe the object as being red.

The perception of color is a purely psychic and subjective event taking place in the inner space of an individual. Objectively speaking, the brightly colored world as we see it does not exist on the outside. The visible world, the colorful world of day-to-day reality, is the product of a transmitter, namely materials objects in the outer space emitting specific electromagnetic waves—and of a receiver, the psychic screen in the inner space. The optical range of what we call reality does not exist on the outside, it exists on the psychic screen inside every individual.

There are similar relations between the transmitter in the outer space and the receiver in the inner space in the acoustic world. The antenna for acoustic signals, (the ear), displays a similarly limited breadth of reception in its function as part of the receiver. Viewed objectively in an analogy to colors, sounds do not exist. What does exist objectively are once again the waves, compressions and expansions of the air similar to waves, that are received by the ear, registered by the tympanic membrane, and transformed into the psychic sensation of sound in the hearing center of the brain—in all its abundance of words, music, and a multitude of other reverberations. The human antennae for acoustic waves, the ears, react to waves in a range between 20 and 20,000 oscillations per second. Slower and faster oscillations filling the outer space are not registered and thus are non-existent in the experience of human reality.

The other aspects of reality made accessible by the remaining three senses, those of taste, smell, and touch, are also created and received by the transmitter in the outer space and the receiver in the inner space, respectively. Just like sounds and colors, there is, objectively speaking, no sensation of touch, smell or taste either physically or chemically.

A sense of taste is triggered by certain molecular structures in food that act as transmitters. Gustatory nerves in the tongue act as antennae that react specifically to these structures and that transmit the impulses produced by the corresponding reactions to the center of taste in the brain.

The transmitter for our sense of smell also consists of molecules—molecules in the shape of vapors with specific structures to which the olfactory nerves in our nose react as an antenna. The signals received by the olfactory nerves—like those absorbed by the gustatory nerves—are transmitted into sensations of smell or taste in the brain in the inner space. We do not know how this transformation of chemical and electrophysical impulses into the psychic dimension of sensation takes place. This is a gaping hole in man's knowledge potential.

The sense of touch, the most ancient, most primitive sense in the evolution of man, reacts to solid objects in the outer world in an unspecific manner. These objects are registered by the corresponding nerves and are sensed as a wide band of sensory observations—ranging from the softest touch to hard resistance—in the inner space by means of certain brain mechanisms.

We can regard every antenna that transmits a sensation of hot, cold, and pain as a specialized tactile nerve. It is obvious that pain does not exist in the outer space; pain is a purely subjective experience in the inner space.

A basic characteristic of our view of reality derived from the above consideration is its inherent limitation. This limitation is defined by the very narrow range with which our receivers react to incoming impulses. How different the world would look if our antennae reacted to electromagnetic waves, and the psychic receiver to another bandwidth in the wave spectrum? For example, longitudinal waves in the wireless range: in that case, we should see into other countries; or to ultrashort X-rays, in which case solid objects would appear transparent. That transparent world would be just as real to us as our present one is now.

Thus we can conclude that the world as we perceive it with our eyes and our other sensory organs represents a reality especially tailored to man, defined by the abilities and limitation of the human senses. Animals see and experience the outer world completely differently with their antennae that react to different kinds and different wave lengths of impulses; they live in a different reality.

Bees, for example, are provided with visual antennae that react to wave lengths in the ultraviolet and ultrared spectrum, they see colors that are invisible to us. Dogs, with a highly developed sensitivity of their olfactory nerves, discover and enjoy smells that are absent in our reality. Bats perceive a reality based on sound by using a sonic radar system.

The metaphor of reality as the product of a transmitter and a receiver clearly illustrates that the seemingly objective picture of the world surrounding us that we call reality is actually a subjective picture. This basic fact signifies that the screen is not in the outer, but in the inner space of every human being. Within himself every human carries his own personal image of reality created by his private receiver.

INSIGHT OUTLOOK

But if every human possesses his own individual picture of the surrounding world, of reality, we have to ask the question; "how true can these personal, individual pictures be?" The answer is that they are all true. They represent the truth, the reality of the respective individuals. In an absolute, objective sense however, these individual realities are not true. There is a transcendental reality, the true essence of which remains a mystery, hidden behind this subjective picture. It is limited by the selectivity, the clearness of modulation of our sensory organs, and the capacity of our mental efficacy—and beyond the manifestation of the physical world that represents our reality. What we know objectively about the physical world, our limited knowledge about what we have called the transmitter, has been revealed be scientific research. *Everything that could be objectively observed in the outer world is matter and energy:* matter, characterized by its chemical and physical properties, in innumerable anorganic shapes and in the shape of uncounted living organisms; and energy as radiation, thermal, and mechanical energy. It has also been discovered that energy and matter can be reciprocally transformed according to Einstein's formula: $E = mc2$ (E stands for energy, m for the smallest unit of matter, and c for the speed of light).

We and the animals of a higher order share the ability—the wonderful ability that defies any further attempt of scientific interpretation—to transform selected energetic and material stimulae from this material world existing on the outside into the physical experience of a shining, living image of the physical world. This image of the physical world we share with these animals does not become a human reality until we additionally include what Teilhard de Chardin called the noosphere of the spiritual world.

The expression sphere, noosphere, evokes the concept of a spiritual atmosphere invisibly flowing around our planet. We must recognize however, that what exists objectively speaking, of the noosphere in the outer space is itself only matter and energy. We can only find the symbols of the spirit in the outer space, mainly sound in the shape of the spoken word and of music, and as matter in the shape of books containing written words, and finally as matter in the shape of human artifacts—paintings, sculptures, architecture, etc. The noosphere, created from the contributions of innumerable individual people during the course of the evolution and history of mankind, exists today exclusively in the form of these material and energetic symbols in the outer space. It only becomes a mental reality in the inner space of the individual thanks to the decoding ability of his individual receiver.

Based on these reflections, the whole interrelation between the outer material world, the transmitter, and the inner spiritual world, the receiver, becomes apparent. Both are necessary as inseparable entities for the formation of what we call reality.

The metaphor of the transmitter/receiver for reality reveals the basic fact that reality is not a concretely defined condition. Rather, it is the result of a continuous input of material and energetic signals from the outer space and their continuous decoding or transformation into psychic experiences in the inner space. Thus reality is a dynamic process being created anew at each moment.

Actual reality only exists in the here and now, at the moment. This explains why a child, living in the given moment much more extensively than an adult, perceives a more real image of the world; it lives in a world permeated with more reality, more truth.

To experience true reality in the moment is one of the main concerns of mysticism. This is where the childlike and mystical experience meet. A poem written by Andreas Gryphius (1616-1664) during the baroque period refers to this:

The years they are not mine
that time has taken from me
The years they are not mine
that still perhaps might be
You moment, you are mine,
and if I treasure thee,
Then He is also mine,
who made time and eternity.

If reality were not the result of continuous changes, but a stationary condition, there would not only be no moments, there would not even be any time, since the sensation of time is only activated by a perception of change. The processional character of reality creates time. Without reality there would be no time, and not the other way around. The transmitter/receiver concept of reality also imparts an insight into the essence of time.

Understanding reality as a product of the transmitter and the receiver however, takes in an especially important meaning when we consider the share each receiver, every individual human, has in the formation of reality. It makes us fully aware of the world-creating power invested in every human. Each individual is the creator of his own world, for it is in him, and only in him, that the world and the abundance of life it contains, that the stars and the sky become real.

Every human being's real freedom and responsibility is founded in this truly cosmogonic ability to create his own world.

Once I have recognized what in reality is objectively on the outside and what is subjectively taking place within myself, then I am more aware of what I can change in my life, where I have a choice, and thus what I am responsible for. Conversely, I know what is beyond my willpower and must be accepted as an inalterable fact. This clarification of my responsibilities is an invaluable help. I have the choice of receiving what I want from the endless programs of the great transmitter; that means I can let the aspects of creation enter into my consciousness—and thus imbue them with reality—that make me happy, or I can let in other ones, ones that depress me. It is I who creates the bright and dark picture of the world. It is I who invests the objects that are, objectively, only shaped matter in the outer world not only with their color, but along with my affection and love, with their meaning as well. This not only applies to the picture of my inanimate surroundings, but also to the living beings, the plants and animals, and to my fellow humans. In one of his poems, Franz Werfel put it this way: "Everything exists if you love! Your friend will become Socrates if you let him."

Just as I am the receiver for the messages of my fellow man, I am in turn a transmitter for him since I am materially located in his outer world. I can only convey my desires, even if they are purely spiritual, an idea or my love to him through what characterizes the transmitter; via matter and energy, namely through my body. Even an unspoken understanding, expressed by a glance or a light touch, even it can only be conveyed by material fingers, material eyes, by the material bodies of the loving couple. Communication would not be possible without matter and energy.

We are reciprocal transmitters and receivers, but even so the image of the transmitter is first formed in the receiver. It is a common experience that different people have a completely different image of one and the same person. Which is the real one? Objectively this cannot be decided as there is no objective image in the outer space. Objectively, the person is nothing other than colorless matter and energy in the outer space.

For myself, as well, my own body is part of the outer world. I can see it as well as experience it with my other senses. In the same manner my sensory organs, the antennae of the receiver ego, are matter and energy and thus part of the outer world. This is not only evident with regard to my eyes and ears; the nerve tract leading from them to the brain are matter, just like the brain itself. The electric currents and impulses that conduct the signals from the outer world to the brain in the nerve tracts and that continue to operate in the brain can be objectified, can be measured as energetic phenomena, and thus should also be ascribed to the transmitter. But then we encounter the great gap in our knowledge I have already mentioned: the transition from a material-energetic occurrence to an immaterial, no longer objectifiable, subjective psychic-spiritual image, to a subjective perceiving and experiencing. This gap in our knowledge simultaneously represents the boundary between the transmitter and the receiver and also where the two fuse and combine into the totality of life.

The transmitter/receiver metaphor of reality would appear to correspond to a dualistic concept of the world: outer space and inner space, objective transmitter and subjective receiver. However, this dualistic aspect dissolves in an all-embracing, transcendental reality if we trace the evolution of human reality, that is the evolution of mankind, back to its source.

So let us start the search for the origin of our corporeal existence, our material side that is, within our metaphor as part of the transmitter.

The inception of our bodies from the combination of the ovum with a sperm cell is sufficiently well known, as its development in the womb, its birth, and its growth based on metabolic processes. But can we really consider the combination of the ovum and the sperm cell to be the actual origin of our material, our corporeal existence? After all, the ovum and sperm do not originate in a void, they come from the parents, and that means there is a transmittal of matter from parents to children. And the parents themselves are created from the ova and sperm cells of their parents, and so on, through uncountable generations. It is obvious that there is a material connection between each human being of our time and all his ancestors—and even further back in evolution, all the way back to the origin of living matter per se, back to the primeval cell.

These considerations indicate that we are, even on a material level, related to our fellow humans and all living organisms, to the plants and animals.

We can continue our quest for the origin and think about the source of the primeval cell. It must have been created by primeval procreation, meaning that the first living cell, the primeval cell, must have been shaped out of inanimate matter, out of atoms and molecules, at the beginning of evolution.

The line between inanimate and animate matter is also the line where scientifically founded thoughts stop and where the realm of imagination and belief begins. This is where we must face the question whether the creation of the primeval cell is based on a coincidence, with a great number of molecules drifting together and combining into a highly organized cell structure, or whether the cell was created according to a plan. We must ask the questions: Was the creation of a coincidental, purely material, or a planned, and thus spiritual event? It seems unimaginable that such a complicated, highly structured and organized entity as a cell could have been created by random chance. It appears obvious—and this is precisely where belief takes over—that the primeval cell was following a plan at its conception. And the primeval cell, for its part, also contains a plan, the plan to reproduce itself, the actual characteristic of life. A plan embodies an idea, an idea in spirit.

In actuality, the atoms themselves, the construction material of the primeval cell, are similarly highly organized entities. They represent a kind of microcosmos that defies being viewed as the product of chance.

It is a remarkable fact that the smallest structural unit of lifeless matter, the atom, and the smallest structural unit of living organisms, the cell, display a similar design. Both are made up of a coat and a nucleus. In both the atom and in the cell the nucleus is the most important component. The characteristic features of matter, mass, and gravity are concentrated in the core of the atom, and the nucleus of the cell contains the basic elements of life, the genetic code and the heredity factors, within its chromosomes.

As the assumption that the origin of highly developed forms such as the atom and the cell cannot be attributed to chance is pivotal to a belief in a spiritual inception and background of the universe, I will try to substantiate it by means of a palpable metaphor. The construction of a cathedral can be used as an example for genesis of an organized form; it would, however, be just as easy to use any number of other examples.

Let us suppose that all the construction material for the building of a cathedral, including the technical appliances and the necessary energy, were readily available at some location. Without the idea of an architect, without his plans and instructions, the cathedral would never be erected.

This kind of consideration must be just as valid for the creation of atoms or for living cells as they are far more complicated and ingenious structures than a cathedral.

If we cannot even imagine the random creation of a cell, the smallest unit of living organisms, then it is all the more difficult to do so in the case of the innumerable more highly developed forms of life in the plant and animal world. Whether evolution took place from primitive to flowering plants, from reptiles to birds to mammals, through gradual mutations or sudden change, is immaterial to these thoughts; nor are the intervals of these events of any importance whatever. *For every new, living organism embodies the realization, the transformation of the plan, of a new idea into reality.*

I would like to use the metaphor of the cathedral once again. Just as the cathedral radiates the idea and the spirit of its architect, every living organism emanates the idea and the spirit of its creator. The more differentiated, complicated, and highly developed the form of a creation is, the greater the spiritual content that can be expressed through it.

Human beings are the most highly developed, differentiated, and complicated organisms of evolution;

this means that human beings express more about their creator than any other creatures. The human brain, with its fourteen billion nerve cells, each of which is connected with six hundred thousand other nerve cells, is the most complicated, highly organized form of life in the known universe. The spiritual element, which even manifests the spirit of its creator in the primeval cell through its idea and its design, has achieved its highest and most elevated development in the human brain. It has attained its momentary perfection in the human spirit, in what we have called the "receiver" in our metaphor. In the human receiver the spiritual abilities have developed to such a degree that it is now capable of being conscious of itself. In man, the most highly developed part of creation, creation becomes conscious of itself.

Within our transmitter/receiver metaphor we can express this as follows: As matter, the human brain is part of the material universe, and thus the brain is part of the transmitter. But the idea and the design of the brain we have developed to the spiritual ability that we have defined as the receiver. In other words, matter and spirit, transmitter and receiver, are fused together in the human brain, and the dualism of transmitter/receiver does not really exist. The transmitter and the receiver are nothing other than mental constructions of our intellect—useful, valuable, and necessary means that are used during the preceding deliberations on a rational understanding of the mechanism on the basis of which human reality is created.

The transmitter/receiver metaphor of reality demonstrates that it is necessary for an idea to be expressed in some form of matter and energy to be existent, to become reality in the outer space. And it shows that every form created in the outer space, from the atom to the living cell, to the innumerable forms of living organisms in the world of plants and animals, from the planets to the suns and galaxies, that every single one of these created forms represents the realization of an idea. To ask about the origin of all these ideas, about the creator-spirit that begat all the created forms and permeates all created forms, is to ask about the origin of all being.

In the gospel according to John we can read: "When all things began the Word already was. The Word dwelt with God, and what God was, the Word was." The translation of "Word" from the original Greek "logos" is controversial. "Logos" could also be translated as "Idea." "When all things began, the Idea already was . . ."

During the past two thousand years, mankind has not developed a deeper understanding of the genesis of the creation than John's. In the preceding discussion we reached the same conclusion based on scientific research and rational thought: a divine idea as the origin and the pillar of creation.

Etymologically, the word "idea" is related to the Greek "eidos" (image, picture). An idea is the spontaneous appearance of an inner picture of something that did not exist previously. The origin of every creative process is an idea. Our ability to have new ideas, that is to be creative, is the gift we share with the creator of the very first, the original idea, the idea out of which the world was born. The gift is our divine inheritance. Our reflections on the essence of reality using the transmitter/receiver metaphor have led up to the primeval questions of being.

At the end of these thoughts on the essence of reality, I would like to mention their value in daily life, the help they can offer for a better understanding of our position as human beings within creation.

As creation exemplifies the material form, the manifestation, the realization of the divine idea, creation—the transmitter of our metaphor—is continuously transmitting the divine idea. Creation contains the message, is the message of its creator to his creations who can receive it, to humankind.

The greatest physician, natural scientist, and philosopher of the Renaissance, Paraclesus, to whom radio and television were unknown, availed himself of a different metaphor to express this fact. He called the creation a book written by the finger of God, a book we need to learn how to read. But instead of studying this book containing the revelation firsthand, we generally adhere to the texts written by human hands. Instead of opening up our senses, our minds to the message of the eternalness of the stars, and the beauty of our earth with all its wonderful creations in the realm of plants and animals, we weld ourselves to our personal problems, encapsulated in a narrow, egotistical view of life. In the process we forget the most important fact: that, due to our corporeal and spiritual existence, we are part of the divine creation and of the all-permeating spirit, and that every single one of us is the "sole heir of the whole world." This truth, encompassing the fact that there are no barriers between the subject and object, between the I and the You, that dualism is a construction of our intellects—this truth is revealed in the course of our reflections, aided by the transmitter/receiver metaphor, on what reality is.

However, truth that is solely the result of a thought process, of rational reflection, is not effective enough to become a decisive factor in our lives. Only when accompanied by an existential, emotional experience does it grow sufficiently strong enough to be able to influence and alter our view of life. An emotional confirmation of truth can be achieved through meditation. Meditation strives to eliminate the subject/object, the I/You barrier to vanquish the dualism.

For this reason the transmitter/receiver concept of reality, which affords us an insight into the origin of the division of the subject and the object and reveals this dualism to be a construction of our intellects, can be a useful object of meditation. Experiencing the lifting of the subject/object dualism emotionally enables us to experience a state of mind called cosmic consciousness or, in the Christian tradition *Unio Mystica*. It can occur as the result of meditation alone or of meditation combined with yoga, breathing techniques, entheogenic drugs, or spontaneously as a gift. It is the visionary experience of a deeper, transmitter/receiver encompassing reality.

Our transmitter/receiver concept of reality can aid us in mentally interpreting this extraordinary state of mind, the *Unio Mystica*.

Initially it reveals to us that the mystical sight is not an illusion, but the revelation of a different aspect of reality.

We only see and experience a small fraction of the surrounding world, of the transmitter, with our everyday consciousness; in a mystical emotional state—when the receiver is adjusted to maximum width of perception—we simultaneously become aware of an endlessly expanded outer and inner universe. The border erected between our ego and the surrounding world by our intellect is dissolved, and the outer and inner space blend into each other. We now experience the endlessness of the outer space in the inner space as well. The unlimited space is now open to an unlimited number of images flowing in, as well as to images from the past, experiences that were collected during the course of a whole lifetime—old images that were stored in the subconscious due to the limited space in consciousness, all these inner images are awakened to new life and fuse with the new ones flowing in. This extremely intensive experiencing of innumerable new and old sensory perceptions and feelings merging inner and outer space creates a sensation of eternity and timelessness, of an everlasting here and now. The body, which in a normal state of mind feels separated from the surrounding world, is now experienced as united with creation, as a part of the universe, which in fact, it is. This also imparts a feeling of security with regard to the *corporeal* existence.

In such an ecstatic condition the transmitter and the receiver, the outer material and the inner spiritual world, the outer and the inner space, are fused together, they are united in the consciousness; thus we must develop a notion of the original idea, the idea that already was, that dwelt with God.

A visionary experience with the intensity of the cosmic consciousness or the *Unio Mystica* is temporally limited. It can last a second, a few minutes, in rare cases a few hours. We are not able to go about our daily business in this extraordinary condition. It is obviously necessary to have a limited ability to perceive and a restricted consciousness to be capable of meeting our daily obligations. To survive in everyday life, we have to fix our concentration on our activities in the environment in which we have to go about performing our given duties. From time to time however, we need a vision, an overview of our life, and an insight into its primeval spiritual reason, so as to observe our place in the universe and our daily obligations and problems from the correct perspective and with the right understanding.

This is why more and more people today are developing the habit of interrupting their daily work and their restless activity to meditate for a few minutes, or an hour, or longer. The goal of this meditation is not to achieve the ultimate visionary experience, the *Unio Mystica,* every time. The goal of this meditation can be to achieve deeper insight into the interrelationship between inner and outer space, between the inner subjective and outer objective world, and to thus become aware of the existence of the transpersonal transmitter and receiver, subject and object, creator and creation, all-encompassing reality. This can fill us with trust, with love, with strength, and with peace.

2

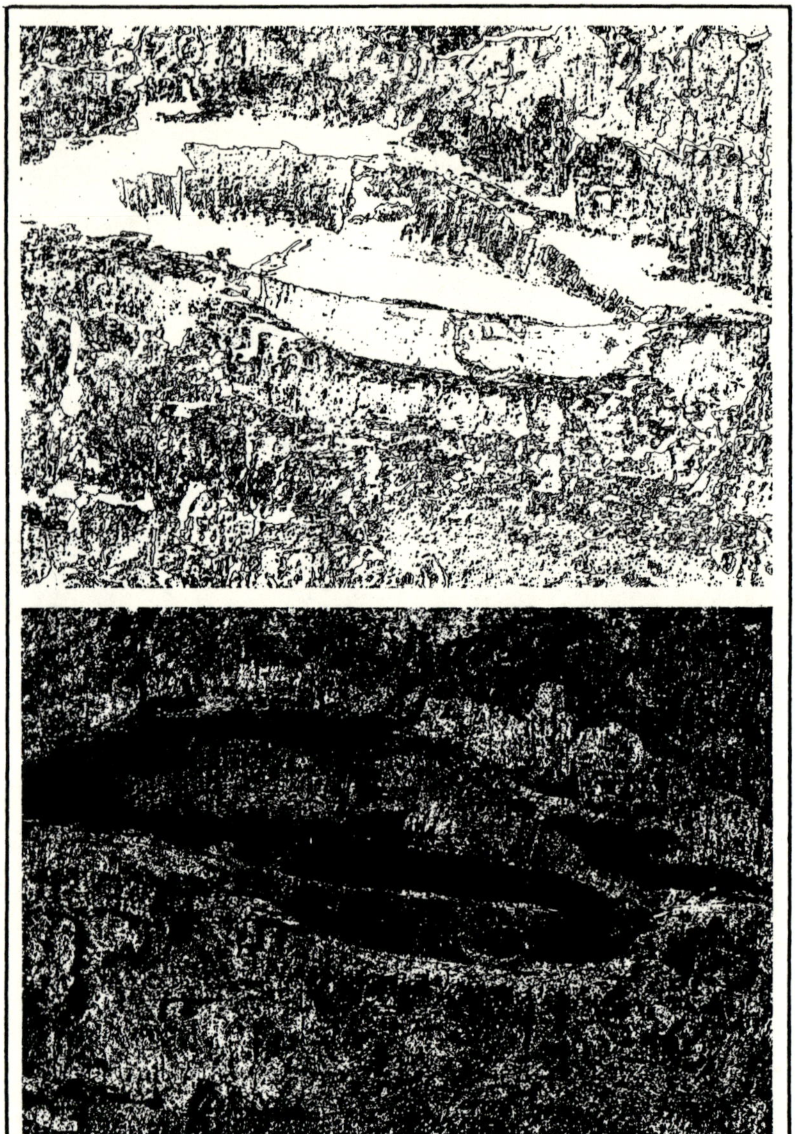

Security in the Natural Scientific-Philosophical View of Life

> Our spirit is growing increasingly aware of the wholeness of the world and our oneness with it due to the progress of the natural sciences. If this recognition of the total unity is not only an intellectual recognition, if it opens up our complete being to a bright omniconsciousness, then it turns into a radiant happiness, into an all-encompassing love.
>
> Rabindranath Tagore
> in "Sadhana" (1861–1941)

Hofmann

> There is no need of proof that an artist reigns in nature. His works may be apparent, but no created spirit gains access to his place of work. We see confirmation of this wherever we look, in every mosquito wing, in every blade of grass, in every snowflake.
>
> Ernst Juenger
> in "The Spanish Moonhorn"

ALL conditions of happiness are based on security in the widest sense of the word. We know the happiness of feeling secure within the house of our parents, within our family, within a friendship. We can also derive a feeling of security that is akin to happiness from belonging to small or large associations, be they of a professional, political, cultural, or religious orientation. In contrast, unhappiness is usually related to being unprotected, separated, lonely, and lost.

This connection between happiness and security does not only apply to the individual fate of a person, but to whole cultural eras. We are talking about the security that a view of life—valid for a certain period of human history and essentially circumscribing his attitude toward life—can offer the human race.

In the following I will try to demonstrate that the securing power of a view of life is mainly based on the relationship of man to creation, and especially to living nature expressed in it. It would appear that the difficulties and seemingly unsolvable problems of the present in spiritual, social, economic, and ecological matters can be traced to a disturbed relationship of man to nature as the common and final cause. The one-sidedly materialistic, natural scientific view of life that is valid in modern western industrialized society is incapable of offering security. It does not express the connection of man to, and indeed his confinement in, living nature. I would like to explain how this could be eliminated through a corresponding expansion and deepening of the natural scientific view of life based on my own experience and personal opinion.

Every cultural group has maintained the memory of a prehistoric time in the shape of myths, the memory of a world in which all humans lived in abundance and blissful security free of all toil and trouble. It was the Golden Age Hesiod spoke of, or the Judeo-Christian tradition of the age of man before the expulsion from Paradise. At that time man was still completely one with creation, was an integral part of it, was secure with it. The world was a garden, the Garden of

Paradise, where all beings lived in harmony and where man found sustenance and everything he needed without hardship and labor.

Whether the people of those prehistoric times were actually as happy as they are described to have been in mythology is unimportant; what is certain is that conditions were already no longer paradisiacal at the time the myths were created, for otherwise their absence would not even have been noted. The ancient authors, to whom we owe thanks for writing down these myths, were already infused with an historical awareness, with the ability to compare the view of life of their times with the view of a past human epoch. This ability, presupposing a critical distance from contemporary events, was already characteristic of a new stage in the development of human awareness.

The biblical allegory of the Fall of man is probably intended to be a description of the entry into this new stage of awareness. The fulfillment of the promise of the Snake: " . . . you will be like the gods knowing both good and evil," split the unity of creation and creature in the human awareness. This new ability to both consciously recognize and differentiate made man the responsible master of his own deeds, but he lost the security that had existed in his subconscious unity with creation. This was the expulsion from Paradise.

Expelled from the abundance proffered by nature in the Garden of Paradise, man, who was no longer protected and now had to rely on himself and on the fruits of his labor, started to build settlements and cities. These are the origins of cultural history which is essentially a history of city cultures. The great cultures grew and died in and with the cities. Wherever on earth there were no cities, time passed by without history.

Cities were places where populations found protection from their enemies and the vagaries of nature. For thousands of years, they were essentially secure place, in which civilizations and cultures could develop. In modern times, the purpose and character of especially large cities has gone through a basic change throughout the world. Centers of life and culture have turned into centers of commerce and industry. Modern urban centers no longer protect their inhabitants from their enemies, they do the opposite; they attract his weapons. And with the noise and the general pollution in the industrial cities, the feeling of security has been lost. But cultural life is still centered in the cities, and world history is still being made in the large cities by people now living there in insecurity and dread. Uncertainty, fear, dissatisfaction, inner emptiness, and aggressiveness are becoming prevalent in our social, political, and cultural life.

Hofmann

Where are the origins of this development that brought about these changes in the habits of mankind, that led to a change of the face of the earth, to the present day view of life, to the present awareness of reality? Temporally they are located in the 17th century, geographically in Europe. At that time, a naturalism completely dedicated to the measurable appeared that was successful in elucidating the physical and chemical laws in the structure of the material world. This knowledge made possible a theretofore unimagined exploitation of nature and its forces. It led to the present worldwide industrialization and technologicalization of almost all areas of life. On the one hand this brought about material affluence and unimaginable comforts in everyday life for a part of mankind, on the other it caused the cited change of cities from centers of life and culture to centers of commerce and industry and a catastrophic destruction of the natural environment.

The reason why it was specifically the European mind that gave birth to the natural sciences, why specifically it was capable of this feat, can probably be explained by the fact that the awareness related to the aforementioned separation of the individual from his environment apparently developed there earlier than in other cultures. For an ego that is capable of contrasting itself with the surrounding world, that can regard the world as an entity, an object, a mind capable of objectifying the surrounding world, was the precondition for the creation of Western, scientific naturalism. This objectifying view of life was already at work in the first documents of natural scientific thought, in the cosmological theories of the pre-Socratic Greek philosophers. This attitude of man toward nature, making a decisive control of nature possible, was first clearly formulated and philosophically substantiated by Descartes in the 17th century.

At the beginning of its modern development, naturalism was still based on a religious view of life. Researchers approached nature as a creation enlivened by the spirit of God. Paraclesus called nature "a book written by the finger of God," the deciphering of which was the task of the naturalist. Kepler recognized the harmony of the world created by God in the laws of the planetary orbits, and none of the authors of the old botanical works forgot to praise the Lord for the miracles of the plant world.

The decisive and consequential change took place when, following the great revolutionizing discoveries of Galilei and Newton, research concentrated more and more on the quantitative, measurable aspects of nature. The qualitative, and encompassing method of observation, for which Goethe spoke out using his theory of colors as an example, was increasingly shunted into the background. The quantitative methods of naturalism that no longer used direct observation required more and more complicated and refined appliances for their measure. They supplied objective results that were substantially independent of the observer, thus additionally supporting the separation of subject and object in the awareness. The disciplines dealing with the measurable aspects of nature, physics and chemistry, experienced incredible growth. Physical and chemical methodologies were incorporated into other areas of the natural sciences, biology, botany, and zoology. As precise sciences, the natural sciences were delineated from the arts, and they were granted preeminence as vehicles of theoretical insights due to the reproducibility and the objectifiability of their results. The great successes of the natural sciences, especially in the fields of physics and chemistry, gave us insights into the macrocosmos and the microcosmos of our world. The practical use of their discoveries and inventions on which the technologies and industries characterizing our age were founded, led the materialistic view of life, its origins being in this naturalism, to its victory. This view of life has become the belief, the myth of our times.

Religious views of life have lost credibility in the general awareness to a concomitant degree. True, ecclesiastical beliefs are still outwardly demonstrated; religious dogmas and ethics are officially still the guidelines in personal and public life. But the spheres of belief and factual knowledge are separated, and practical life is determined by the latter. Even if a statesman swears on the Bible, he still only trusts in the reality of the atomic bomb and makes his international political decisions accordingly. Only the world created and controlled by technology is viewed as real today, i.e. as being important to everyday life, the extent of which can be seen in the fact that environmental protectionists, who see our true real home in the original nature and who believe in its powers, are still considered to be somewhat quaint.

The attempt at a short explanation given above of how the present world situation came about could be summarized by once again citing the biblical example of the Fall of man.

After having been expelled from the security of the Garden of Paradise into insecurity and self-responsibility, man, endowed with an expanded ability to perceive, was empowered to dispose over the earth and its treasures, i.e. "Subject the Earth to yourself." But instead of turning his new home into an earthly Garden of Eden to find new security, man, misunderstanding the divine directive, destroyed the earth while abusing his newly achieved mental ability and is now about the make it completely uninhabitable.

Does the development have to continue in this direction and does the destruction of the inner and outer world have to keep on expanding? Pessimistic forecasts are multiplying. What is sure is there is no reverse development, only forward development, a further development of the history of ideas that has been achieved, of the present state of awareness and that accompanying natural scientific view of life is possible. It is just as impossible to reverse the expansion of technological-industrial civilization. All that can be done is to give its further development a new goal, a new meaning.

A precondition and basis for a change to the better would have to be the healing of the "European fate neurosis" as Gottfried Benn has called the split perception of reality. It would be necessary to revive an image of reality in the general consciousness in which the individual no longer experiences himself as separated from the surrounding world, but as one with creation.

It is essential to recognize that the one-sided belief in the natural scientific view of life is based on a momentous error. Certainly, everything it contains is true, but this only represents half of reality; only its material, quantifiable part. All of the spiritual dimensions that cannot be described in physical or chemical terms, which include the most important characteristics of that which is living, are absent. These need to be integrated into the natural scientific view of life as a supplementing half, so that an image of a complete, living reality is created that also includes man and his spirituality. Consciously experiencing this complete reality lifts the separation between the individual and the surrounding world, between man and creation. This would heal us of the "European fate neurosis." Such a natural scientific view of life, supplemented by the dimensions of that which is living and enhanced by meditation would be able to offer us security again.

The goal is not to deny the validity of the natural scientific view of life and to downplay the value of the measuring sciences. We are only talking about recognizing their titanic myopia. To the contrary, we are making the argument here that the natural scientific view of life is the only solid, stable basis on which to build, and on which we must therefore continue to build spiritually as well as materially. The enormous amount of substantial knowledge, the deep insights into the material construction of the world, the earth and its living organisms are undeniably great achievements that cannot be neglected. The expansion of our awareness of reality they caused cannot be annulled, for they represent a new, higher level in the history of the development of the human spirit.

In the following I would like to describe how my view of life was influenced by my knowledge and my insights as a natural scientist. As the following observations are thus mainly personal opinions and views, so that the subjective is an important factor in them, a few remarks about the subject, about myself, would appear to be indicated in introduction.

As a boy I had frequent mystical experiences of nature while roaming through the fields and forests. A field of flowers, a sun-brightened spot in a forest, some place in my surroundings would suddenly appear in strange clarity. It was as though the trees, the flowers now wanted to disclose their true essence to me, and I felt connected to them in an indescribable feeling of bliss. These experiences impressed themselves on me deeply even though they were usually only of very short duration. It was they who not only kindled my love of the world of plants, they defined my whole view of life in its basic outlines by disclosing to me the existence of an all-encompassing, securing, deeply gratifying reality hidden from everyday life.

This interest in the problem of reality, primarily displaying itself as a material reality, was the reason I decided to study chemistry even though my classical education in Latin was considered to be the basis for work in the humanities. An additional factor in selecting the field of chemistry was my desire to find some stability in a hard, irrefutable area of knowledge. In philosophy, history, literature etc., opinions are contradicted by opinions, convictions by convictions, for all systems of the mind can be discussed. In contrast, the material world is irrefutable and its inherent laws are fixed. The science that offers us insights into this tangible, hard, but basically still mysterious part of our world, into matter, is chemistry.

Chemistry is generally regarded as the most materialistic science. However, it is only the object of chemistry, matter, that is materialistic or material; its scientific-methodical research is, like all scientific research, of a mental nature.

At this point I would like to insert a digression about the image of the natural sciences and especially of chemistry, in the general awareness. Superficial knowledge has brought about an incorrect impression of the essence and the meaning of the natural sciences. Today our views and opinions are mainly determined by the mass media, uniformly and worldwide. What they pass on as knowledge—it is called information today—is usually only partially correct, superficial, and not aimed at truth and verity, but mainly at sensationalism. Programs have to sell well. For example, what the average person thinks about chemistry has, to a great extent, nothing whatever to do with chemistry as a science. The cliche of the chemist is that of the man in the white coat wearing glasses and mixing up something in a test tube. He is the poison mixer par excellence. This concept already demonstrates the existing erroneous understanding of the essence of chemistry. The mixer of poisons would be a physicist, not a chemist, for mixing is a physical process. Chemistry only starts where the *transformation* of substances, of matter, is the issue. Beyond that, the common perception of the expression "chemistry" is exhausted with the image of the chemical industry, and the smell of environmental pollution connected with it. Only a small minority of the population is aware of the importance of the theoretical insights of chemistry as a science of the construction of the whole visible, material world.

So much for my digression on the incorrect perception of the essence of chemistry; it applies to the other natural sciences as well. I thought it was necessary because it demonstrates a superficial knowledge that is mainly at fault for the incorrect evaluation of the natural scientific view of life.

Studying chemistry fulfilled my expectations. It opened a view into the inside, into the visible construction of the visible world: into the molecular and atomic structures and the microcosmos of the atom. I learned that the realm of minerals, the worlds of plants and animals, including man, consist of the same few elements. Of the total 92 known atoms, the by far greater number only exist in traces. There are only a dozen elements that are decisively involved in the construction of the earth and its biosphere; hydrogen, oxygen, nitrogen, silicon, calcium, sitrontium, phosphorus, sulphur, iron, nickel, magnesem sodium, potassiun, to only name the most important. If we trace the common construction elements of the atoms even farther back to the protons and neutrons that form the atomic nucleus and the electrons circling the nucleus, the number of construction elements of the whole world is reduced to three.

The reduction of the world to a few dead elements as its last reality has been used as the basis of a materialistic view of life. This expresses an incredible exaggeration of the role of matter in creation. It is nothing less than reducing the miracle of a cathedral to the number and the quality of the building stones used; ignoring its blueprint, its beauty, its meaning, and consequently finding no reason to think about an architect. In addition, the cathedral lacks the dimension of life, so that the comparison does not even express the complete magnitude of the inadmissible reduction of the essence of creation to the level of chemistry.

It is hard to understand why chemists specifically—who should know what is within the powers of chemistry and what its limitations are—are not increasingly attacking this materialistic view of life reduced to the level of chemistry. In fact it is more the biologists who have too great a trust in chemistry, and who are trying, within their rational aspirations, to trace the phenomena of life to chemical reactions.

This is one of the essential points of my work. I want to demonstrate that opinions diverge at the point of different perceptions of the role that chemistry plays in the natural scientific view of life. On the one side we have the role of chemistry and its laws as the final cause and reason for the creation of the visible world, on the other the role of chemistry as the science about the construction material being used by a spiritual force to construct creation in its colorful abundance.

I would now like to present a few thoughts that show how, in my case, it was above all my knowledge as a chemist that disclosed a natural scientific view of life to me that gives me security.

When, in the garden or while taking a walk, I stop to meditate on a plant, I not only see what a non-chemist sees, its shape, its color, its beauty; thoughts about its inner construction, its inner life and the chemical and physical process they are based on also force their way into my mind. The plant is composed of innumerable individual chemical compounds. I can visualize their chemical formulas. To name only a few: the composition of the skeleton substance, the cellulose made of saccharose rests, then the complex formula of the leaf green chlorophyll, consisting of several nitrogenated hydrocarbon rings with a central magnesium atom, furthermore the structural formula of the colors in the blossom, for example the formula of a blue color, of a anthocyanin. Most of these plant components can also be produced artificially, by means of chemical synthesis. I know the efforts necessary for this in a laboratory, the construction out of reactive groups of atoms via many intermediary steps, at high or low temperatures depending on the type of respective chemical reaction, sometimes in a vacuum, at times under high pressure etc. Chemist Hans Fischer Munich, who together with a whole school of assistants and students, supplied us with the main body of work for explaining the structure of chlorophyll, was rewarded with a Nobel prize. Harvard Professor Robert Woodward, who finally succeeded in synthesizing chlorophyll completely, was also honored with a Nobel prize. My respected teacher and doctoral advisor, Professor Paul Karrer, who worked on explaining the structure and synthesis of blossom colors, on the anthocyanins and the carotenoids during the twenties and thirties, was also awarded a Nobel prize for his work. And all of these achievements were only possible on the basis of the knowledge acquired by chemists of preceding generations. I am elaborating in such detail to show the enormous chemical effort hidden behind the synthesis of each of the numerous substances of which a plant is composed.

And every little blade of grass is capable of this effort. It produces these materials—the work of hundreds of chemists for many years would not be sufficient to synthesize them—quietly and humbly with light as its only source of energy. A chemist has to marvel at this.

But nevertheless, it is chemistry, and we know its laws today. We can imitate them even if it is incredibly difficult and means we have to tap all our possibilities.

But while observing the plant I am concentrating on at this moment, other thoughts as well force their way into my mind. They relate to the manner *in which chemistry is made subservient,* something that cannot be explained, it can only be described. Place and time enter into the equation, things that have nothing to do with chemistry anymore. After all, each of the uncountable synthetic processes has to take place at a given time and a given location so that the predetermined outer shape of the inner construction of the plant, its different organs with their various functions can be formed. This adds a multitude of physical processes and forces, such as diffusion, absorption, capillary phenomena, to chemistry. All this is unimaginable without a blueprint and without a coordinating power.

Cell physiology and molecular biology offer an explanation for this. The blueprint is pre-programmed in the chromosome map of the nucleus of the cell. It is written down by means of the four letters of the genetic code, by means of the four different DNA-molecules.

These are incredible scientific insights into a wonderful mechanism. It is important, however, that we comprehend that this only explains the mechanism; we know the four letters of the biological alphabet. The decisive question about the origin of the text remains unanswered. In addition, we have to take into consideration that chemical structures, such as the one represented by the nucleic acid group of DNA, can, by their very nature, only direct the chemism of an organism, they cannot determine its shape.

In conclusion, I would like to talk about a third kind of thought process that goes through my mind as a chemist while meditating in the garden or while walking through the woods. It revolves around the relationship in the chemical structure of human and plant organisms and the inclusion of man in the biocosmos expressed in it.

Every higher organism, be it plant, animal, or human being, stems from a single cell, the fertilized ovum. Cells are the smallest units of life of which organisms are constructed. Plant, animal, and human cells not only display a similar structure—consisting of the nuclei protecting the chromosomes,

Hofmann

the former embedded in the protoplasm and the whole contained by the cell membrane—they also have a largely identical chemical composition. In spite of the innumerable variations in the chemical structure of the various organ parts and tissue types, in their totality the same classes of organic chemical compounds are involved in the material composition of the bodies of humans and animals as in those of plants. In both the worlds of plants and animals these are proteins, carbohydrates, fats, phosphatides, etc., they themselves being composed of the same simple structural units, the amino acids, sugars, fatty aids, etc. that are the main elements forming the material basis of organisms.

This similarity of material composition exists in relationship to the great metabolic and energetic cycle of all things living in which the realm of the plants, animals, and humans are united. The energy that keeps this cycle of life going is supplied by the sun. Primarily it is nuclear energy created by the transformation of matter into radiation energy during nuclear fusion. The daystar transmits this energy to the earth in the shape of light. The plant, the green carpet, the vegetable kingdom, is capable of absorbing this immaterial flow of energy in maternal susceptibility and of storing it in the shape of chemically bound energy. During this process, the plant transforms organic matter, water, and carbonic acid onto organic substances with the help of the leaf green chlorophyll as the catalyst and light as the energy source. This process, called the assimilation of carbonic acid, supplies the organic construction units—sugars, carbohydrates, amino acids, proteins, etc.—for the assembly of the plant, and thus of animal organisms, as well. *Energetically, all life processes are based on this light absorption by the plant.* When the nutrients stemming from plants are combusted in human organisms to obtain the energy necessary for the life processes, the reversed process of assimilation is taking place: the organic nutrients are transformed back into anorganic matter, into water and carbonic acid, while releasing the same amount of energy originally absorbed in the form of light. Even the thought process of the human being is supported by this energy, *so that the human spirit, or consciousness, thus represents the highest, the most sublime energetic stage of the transformation of light.*

I have taken the liberty of recapitulating these basic insights of the natural sciences, facts that can be looked up in any elementary text book on biology, specifically because they are no longer given appropriate attention due to their general familiarity. They are part of a body of knowledge that only receives consideration on a strictly intellectual level. Moon landings, space travel, science fiction books and movies, in which living nature is not even shown any more, have a greater influence on the minds and imaginations of the inhabitants of our industrial society and define their values of life and perceptions of reality.

However, to someone with close ties to nature, to someone who, through meditation, lets these natural scientific findings some to conscious life, the tree or the flower he is contemplating no longer appears simply in its objective beauty. Rather, he feels deeply connected to it as a fellow creature, as a living being created by light.

I am not talking about some sentimental enthusiasm for nature, about a "back to nature" in Rousseau's meaning. Indeed, the roots of that romantic movement—with its search for an idyll in nature—can be found in man's feeling of being separated from nature.

What I have tried to describe using the example of our relationship to the world of plants is the elementary observation of the actually extant unity of every living thing, the development of an awareness of security in a common basis of creation. The more original flora and fauna of earth have to give way to a dead technological environment, the rarer this kind of exulting experience will become.

Those important experiences of my youth I mentioned earlier, where field and forest would suddenly

INSIGHT OUTLOOK

appear in the inexplicable light of enchantment, have nothing to do with sentimentality. As I know today, it was actually the light of the reality of being in the common basis of life together with the plants that triggered this enchantment in the open mind of a child.

In the above I have tried to show that, from the standpoint of a chemist, the insights of natural scientific research need not lead to a materialistic view of life. Quite the opposite is true: if they are contemplated and understood correctly, they invariably point to an altogether inexplicable, spiritual primordial basis of creation, to the miracle, the mystery—in the microcosmos of the atom, in the macrocosmos of a spiral nebula, in the seed of a plant, in the body and spirit of a human—to the divine.

Meditative observation starts at the depth of objective reality to which perspicuous knowledge and insight have penetrated. Thus meditation does not imply a turning away from objective reality. The opposite is true: it means you penetrate with a deeper and broader understanding. It is not an escape into mysticism, but a search for an encompassing truth by simultaneously observing the surface and the depth of objective reality stereoscopically.

By observing natural scientific discoveries through a perception deepened by meditation, we can develop a new awareness of reality. This awareness could become the bedrock of a spirituality that is not based on the dogmas of a given religion, but on insights into a higher and deeper meaning. I am referring to the ability to recognize, to read, and to understand the *firsthand revelations* "in the book written by the finger of God," as Paracelsus designated creation.

Thus it is necessary to recognize the laws of nature uncovered by the research of the natural sciences for what they are, to see that they are not primarily the instructions and means for the exploitation of nature, but revelations of the metaphysical blueprint of creation. They reveal the unity of all things living in a common spiritual primordial basis.

Another important insight affecting the position of man in creation can be derived from the hierarchic structure of all things being, structures explained by natural scientific research. It is the hierarchy in the structure of the anorganic, from the elementary particles to the atom, molecules, rocks, planets, suns, and on to the galaxies, as well as in the realm of living matter, from the cells to tissue, organs, organ systems, and all the way to complete organisms. Thus the dual function of all things being becomes clear: on the one hand it is as an independent whole, and on the other as part of a higher order. So as to be able to meet their obligations as a part of a higher order, all units are inhabited by the desire and the strength to achieve their own completeness. This is where the obligation of every individual human being to work on himself shows as a law of nature, and thus as a metaphysical revelation—the obligation to perfect his given abilities and to expand his knowledge, and thereby his awareness, to be able to do justice to his destiny and his duty as a spiritual being participating in creation.

If bliss encompasses the final goal of this destiny—as Thomas Aquinas formulated it; *ultima finis vitae humanae beatitido est*—and happiness presupposes security, then, the preceding development of the human race could be understood to mean that we are to develop out of the shadowlike, mythic happiness of our security in a dreamlike state of being into the happiness of a completely aware, glowing existence of freedom and responsibility.

It is true that we have achieved a high degree of awareness and freedom today thanks to the insights of natural scientific research and their technical applications. Now, however, it is important to once again become aware of the security in creation we lost as the precondition for all true happiness; it is important to once again recognize what man overlooked in his titanic arrogance; that we are rooted and secure in our common creative primordial basis of all things living.

If this insight were to enter into our collective consciousness, the result would be that natural scientific research and the hitherto destroyers of nature, technology and industry, would be applied to transform our world back into what it once was—into an earthly Garden of Eden.

In place of utopian projects of space travel, of insane arms programs, and senseless contests for economic and military supremacy, this could then become a goal of humanity, uniting the peoples of the earth and promising true happiness. We could derive new, sensible standards from this goal, pointing the way to a solution to all of today's confusing economic, social, and cultural problems.

3

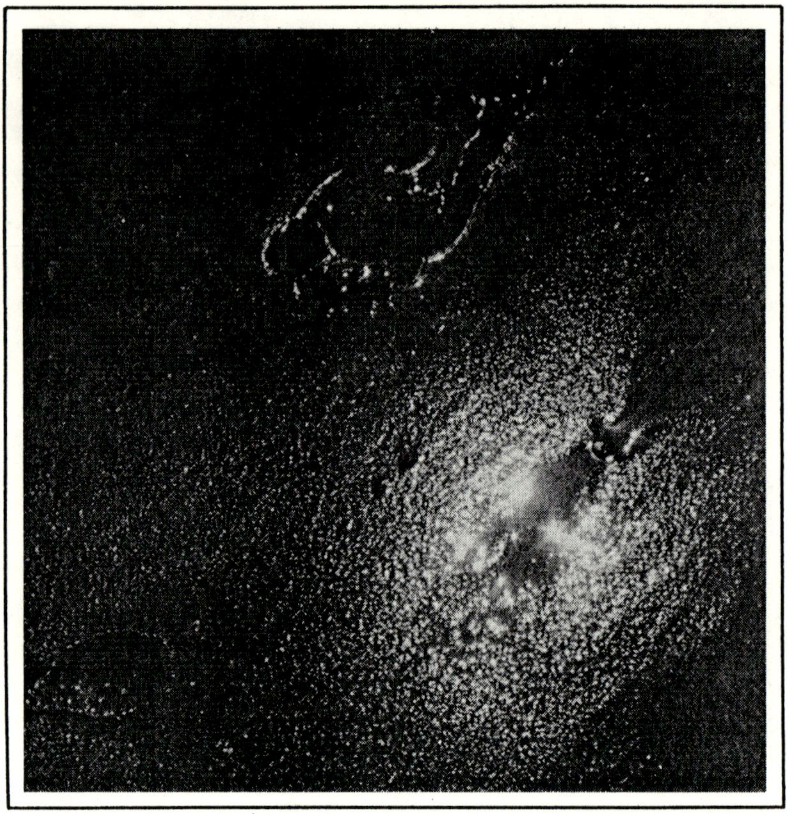

On Possession

> You will not be able to take joy in the world until you feel the ocean flowing in your veins, until you clothe yourself with the heavens and crown yourself with the stars, and see yourself as the sole heir of the whole world—and more than that, for there are people living on it who, like you, are the sole heirs.
>
> Thomas Traherne
> (1638–1674)
> in "Centuries of Meditation"

IT is amusing and informative to ponder the original meaning of words. They derive from a direct experience of reality and relate to elementary facts and activities of our existence. Thus they possess, from the

59

time of their inception, a figurative character that in the course of time has been worried away like the image on a coin, so that it is only visible when studied closely.

Using the word "possessions" as an example makes this transformation very vivid. The corresponding verb to "possess" relate to the process of "sitting down on something." I pos-sess a chair originally meant, I am sitting down on a chair. This turns it into my possession. It has become my chair, if not in a legal sense then at least to the extent that it is my chair in contrast to other chairs on which other people are sitting.

In very early human communities, when the word was created, possession probably designated only what one could use personally. The main thing one pos-sessed was a horse. It, along with the other objects used in daily life, was what constituted a possession among people still leading a nomadic life. Since then, possession and possess have achieved a much more encompassing as well as symbolic meaning. Since the introduction of the legal concept of property as the judicial recognition and lawful protection of a possession, it has become possible to acquire more property than it is possible to possess or, in the original meaning, to use personally.

This notion planted the seed of an important part of the human tragedy. As the word property also implies the concept of the right to dispose over a possession and thus also translates into power, the acquisition of property simultaneously leads to an accumulation of power. The striving for power, the attainment of power, the use of power in a positive sense, and the abuse of power are fate determining fact in our personal lives and in world political events.

The relationship between property and power is the reason for the abolishment of private personal property in communist states. This increased the property of the state, and its power grew accordingly. In capitalist countries power is effectively wielded by groups that have accumulated enormous amounts of property.

Power based on property has little to do with human contentedness; it is more inclined to detract from it. For this reason this essay will concentrate less on possessions with the character of property, and therefore on their relationship to power, and more on possessions in their original meaning, on their existential importance to the individual. Legally, possession is defined as the actual power of a person over an entity, meaning that he can do whatever he wants to with the corresponding object, that he can dispose of it in any way that he pleases. I can also possess something that is not my property; if I have a borrowed or some tool in my workshop and use it in any manner I like, the tool is in my possession, but it is not my property. The opposite is also possible; I can call something my property that I do not or cannot even possess, if we understand possess to always indicate some kind of use in the widest meaning of the word, as an active or receptive relationship to an object.

Property does not become a possession until there is an existential relationship between the owner and the property. A possession does not become property until there is an abstract relationship, a legal attribution.

The fact that there is an accompanying verb to possession, possess, but no such verb for property, additionally underlines the fundamental difference between the two.

Many fruitless efforts, many disputes, much discontent would evaporate—and there would be a corresponding increase in equanimity, cheerfulness, and happiness—if, generally aware of this difference, we were to concentrate more on real possessions and less on property. A Chinese aphorism underlines the meaning of this in the most succinct way: " The master said: 'My garden'. . . and his gardener smiled."

INSIGHT OUTLOOK

The master is correct in telling his friends that it is his garden, for it is his property. But it could be that he is hardly ever there. Or perhaps he does walk through on occasion to show his visitors an especially beautiful plant and the newest pavilion. In contrast, the garden is the natural element of the gardener. He lives in and with it. He planted the trees, he prepared the flower beds, he knows every single flower, every single plant. He cares for them lovingly, he watches them grow, blossom, and die. He knows the garden in the freshness of the morning dew, he walks among the flower beds one last time at nightfall when the aroma of some of the flowers is especially pervasive, and during the heat of the afternoon he enjoys taking his nap in the pavilion. He loves the garden with all his heart. It also he who "possesses" the garden from dawn to dusk; he is its real possessor. It is *his* garden, and that is why he smiles when his master says: "My garden . . ."

In the above example of the master and his gardener, the proprietor at least still had the possibility to enjoy his garden. But if we look at large landed properties the difference between the proprietor and the possessor is even more obvious. It is not necessary to be the possessor of meadows, fields, and woods you strode through to be able to take pleasure in the flowers along the way, the winds playing in the trees, or the other sights and sounds encountered along the way.

63

The woods around where I am fortunate enough to live are the property of the surrounding communities and partially of a private foundation. During my long, almost daily walks through the woods it is rare that I encounter someone else, and I never run into the communities or the foundation. I experience the forest with the birds, the deer, and all the other animals that live there. And when, on occasion, I do meet a lone hiker, our greetings are almost always accompanied by the exchange of a few friendly words vibrant with the sympathy of two people who are both aware that the other is the possessor of these woods.

There is an old boundary stone at the edge of the forest near the country border. On one side it displays the coat of arms of the neighboring monastery of Mariastein, for several centuries the owner of the forest glade on which our house is located. On the other side, looking out over France, a bas-relief still clearly shows the arms belonging to none less than the great French statesman Jules Mazarin (1602–1661). In recognition of his outstanding services during the so-called Pyrenees Peace Accord between France and Spain, Louis XIV presented him with the country of Pfirt and other adjacent areas in the Sundgau. Considered to have been one of the wealthiest people in Europe, this avaricious statesman died without ever having set foot on his Alsatian possessions.

This clearly exemplifies the illusory character of this kind of ownership: it is only a property, not an actual possession. The vagabond drifting through beautiful countryside was, in reality, the possessor of the land; the rich man in Paris only owned it on a piece of paper. At this point the objection could be raised that, viewed from a different angle, the value of Mazarin's Alsatian possessions was not illusory, that it was indeed very concrete: he collected money from them in the shape of taxes and other duties.

So now we have to consider the possession of money. If true possession means a corporeal, sensory relationship to an object, then money can never become something I possess; it will always be a symbol of possession. Money is, quite understandably, an especially sought after property, as it can enable us to obtain many things that we can use, apply, enjoy; it can give us real possession.

There is no need to list all the things money can buy. The universal possibility of converting money into many different kinds of possessions gives it an especially manifold power that is inherent in property. However, it is useful—and if you have little money comforting—to realize where the possibility of converting money into possessions reaches its limits.

In a case where the value of a possession is solely based on consumption, on gratification, it is determined by the owner's capacity for gratification. Even a millionaire can only eat as much as his stomach allows. If he orders more, he will have to leave it on the table. What holds for food is even more evident in the case of liquor. Here a price is exacted for transgression: a hangover, an alcohol poisoning that can be fatal.

A rich person can, however, make the gratification of his bodily needs and pleasures more enjoyable than can a poor one; but this only applies to a limited degree. If you can spend more money, say for a meal, the pleasure of eating will indeed be more increased. But the simplest meal tastes better to someone who is hungry than the most refined one does to someone who is not. In general, the rule holds that the intensity of gratification of corporeal pleasures is defined by the extent of the corresponding need, by the appetite in the widest meaning of the word. But an appetite cannot be purchased. This compensates for many social injustices.

But the grand compensation is that every human is in possession of the ability to be a possessor. A possessor-possession relationship is only possible between a subject capable of perceiving an object and its usufruct, whereby object should also be understood to include spiritual contents and usufruct a relationship consisting of love and joy. As every human, and only the individual human, is capable of perceiving and loving, he alone can take possession of objects in the outer world. This ability not only permits him to possess individual items in the outer world in the manner I explained beforehand, he can also be, in the true meaning of the word, the possessor of the whole world. This is the divine gift placed in every human's cradle. Most of the time, however, our view is obstructed by things in our direct vicinity, our thoughts are occupied by personal interests and problems, so that we fail to see the miracle and the beauty of creation as a whole. The heavens and earth, the sun and moon, walks through the field and forests during the changing seasons, have become matters of course and are hardly noticed anymore.

Nor do we think about the fact that the colorful, sensually vibrant world as we see and experience it is created in ourselves.

The first chapter in this book extensively discusses this wonderful occurrence, the interrelation between matter and energy in the outer space as the transmitter and the spiritual center that makes us aware in the inner space of every human as the receiver, out of which reality is created.

There is only one outer, physical space that I share with all other humans; in contrast, I am the sole possessor of my spiritual inner space. That, and no other, is the place where the image is created we call our reality. This image has grown in me by means of my senses. It belongs to me. I am the sole possessor of this image that is identical with the world, with my world.

This is what Thomas Traherne means in his motto that precedes this essay, with his invitation to regard myself as the sole heir of the whole world. Every human is, in fact, the sole possessor of the whole world, including his fellow men who are a part of that world, for the world only becomes a reality in one ego, in every ego.

This knowledge, derived from natural scientific insights, namely that the whole world is in my possession, is not in itself sufficient for me to be able to take joy in this world. What has to come to pass is what Traherne means when he says; "I have to feel the oceans flowing in my veins, I have to clothe myself with the heavens and crown myself with the stars." Rational knowledge has to be augmented by emotional experience. I must not remain separated from the oceans, the heavens, the stars. I must feel that I am within the creation and the creation is within me, that we are one. Then the world belongs to me, as I belong to it. Not until then will my heart recognize its real beauty, will I feel secure and be able to take joy in it.

4

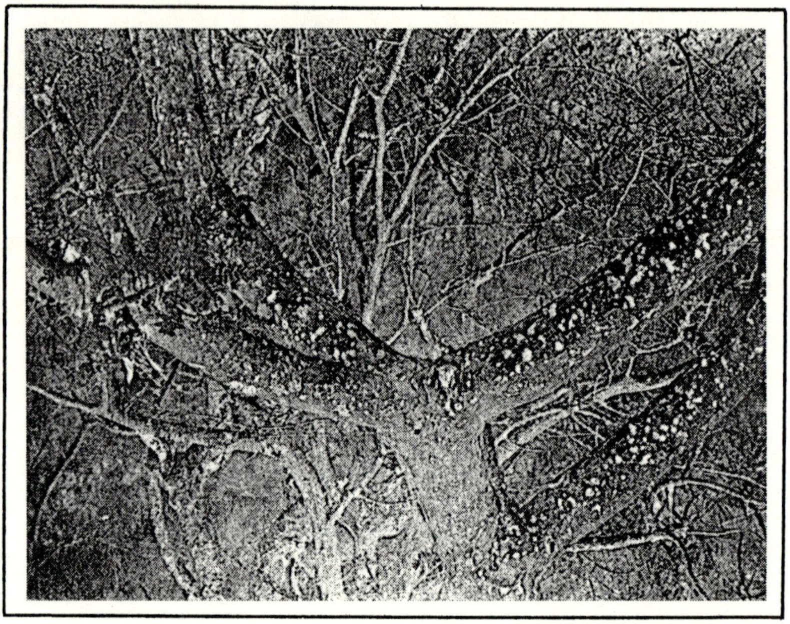

Botanical Reflections on the Death of the Forests

IN spite of the public discussion about the death of the forests, two fundamental botanical reflections are rarely if ever mentioned, even though they are more than obvious.

One relates to the question of why air pollution has a negative impact on the plant world, on the trees of the forests, before it affects the world of animals and man. After all, generally one would consider a fir or beech tree to be more robust and less sensitive than an animal or a human.

But the enhanced sensitivity of plants with regard to pollutants in the air is immediately apparent if we consider the fundamental differences in the biology of plants and animals.

We "only" need air for oxygen, which we use to combust our food to obtain energy for our life processes. A plant, on the other hand, derives the main bulk of its food from the air, taking carbon out of it in the shape of carbonic acid (specifically: carbonic anhydride = carbon monoxide = CO2). As air only contains 0.035% carbonic acid, in contrast to an oxygen content of 21%, a plant has to come into contact with an incomparably large amount of air to meet its large carbonic acid requirements than is necessary for a human to inhale in order to obtain his relatively smaller quantity of oxygen. To this end, the green tissues of plants, the leaves and needles, in which the assimilation process of the carbonic acid takes place are embedded with a highly developed ventilation system that enables them to filter the greatly diluted carbonic acid out of the air. The air can penetrate into the interior of the leaf or the needle via fine pores, so-called stomata, whereby energy single oak or beech leaf has more than half a million of these.

Necessary for metabolism, this extensive, highly intensive ventilation of a plant explains why so many more air pollutants (sulphur dioxide, nitric oxides, ozone, lead, dust, and others) are trapped in them than in animal organisms, so that the effects of a poisoned environment will first be noticed among plants, than among humans and animals.

INSIGHT OUTLOOK

The other thought not taken into consideration about the death of the forests relates to the question of why, in the realm of plants, it is specifically the trees of the forest that are the victims of the poisonous substances in the atmosphere. As far as we know, there is as yet no sound explanation for this. There is a possibly frightful danger lurking behind this lack of knowledge. For if there is no known basic difference in the assimilation mechanism of carbonic acid in forest trees and fruit trees, or in other useful plants such as potatoes, wheat, etc., *we have to consider the possibility that the plants man uses for food will start to die in the foreseeable future, as well.*

To review briefly, the plants constructs its organism, consisting of carbonic compounds, from the carbonic acid in the air and from hydrogen, using the light of the sun as its source of energy and the green of the leaves (chlorophyll) as its catalyst, in a process called the assimilation of carbonic acid, or photosynthesis. The hydrogen is obtained by photochemically splitting the water rising in the roots. The oxygen released during the process is passed into the air through the stomata.

In our organism and in that of all animals, the exactly reverse process takes place. The organic substance synthesized by the plant, our food, is combusted while absorbing oxygen; at the same time, we obtain the energy absorbed by the plant in the shape of life and transfer the combustion products, carbonic acid and water, into the atmosphere by exhaling. The cycle is thus complete.

Aside from the basic carbohydrate cycle, there are other cycles in which nitrogen and minerals play a role; they, too, are powered by the energy of the sun.

When looking at photosynthesis, we are confronted by the basic process of creation that supports all life on earth be transforming the *immaterial* light flow from the sun into the *material* energy of plant organisms by means of the green carpet of plants of the earth, these organisms for their part being the life basis of the world of animals and humans. The death of trees that can be traced to a disruption of the photosynthesis due to damage caused to the green plant cells by pollutants in the air is the harbinger of a threatening interruption of the basic process in our own life cycle.

The basic concepts of the assimilation of carbonic acid, of photosynthesis, are described in every elementary textbook of botany. Unfortunately, however, it is specifically this kind of fundamental knowledge about the basis of our lives that, having no practical use, is frequently shelved along with the textbooks. But today it is of paramount importance that we all recall to memory these natural scientific insights; for they make us aware that the death of the forests is beginning to acutely endanger *the basis of all life on our planet,* and that the postponement of possible measures to banish this impending catastrophe would not only be boundlessly irresponsible, it would be a crime threatening all of life.

5

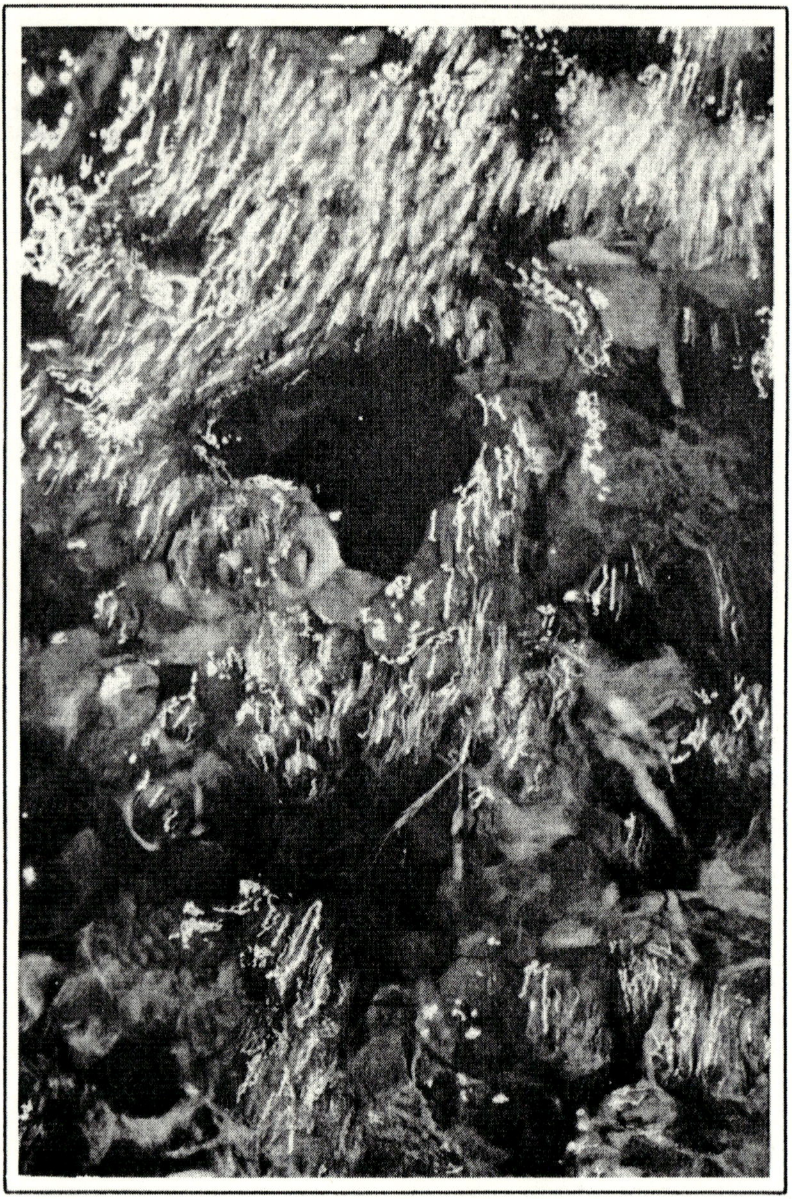

The Sun, a Nuclear Power Plant

IF we take a look at the passionate worldwide discussion surrounding atomic power plants, we could reach the conclusion that essentially, the problem of exploiting nuclear energy simply revolves around answering the two following questions:

a) Will future energy needs be so large that we will need atomic power plants?

b) Is the operation of atomic power plants so safe, and is the problem of atomic waste so solvable, that we need have no fear of catastrophes or hereditary biological damages to the human species?

Both are questions that can only be answered by specialists and competent scientists—if we can answer them based on the given facts and knowledge we have today.

But the scientific experts disagree on the answers to both questions a) and b). Thus if we look at the matter from these two standpoints only, it is unclear whether we should or should not agree to the construction of atomic power plants.

There are, however, other reflections related to the problems of the use of atomic energy that have nothing to do with the answers to questions a) and b), and every thinking person can ponder them without having to rely on the experts or specialists.

I am referring to the thoughts and considerations that surface when we contemplate *the fact that the sun is nothing other than a gigantic power plant.*

Our knowledge of the chemical and physical processes taking place on the sun is quite precise. They are all nuclear reactions. Among them, the fusion of hydrogen nuclei into helium nuclei is of great importance. Together with these processes, incredible quantities of energy are radiated out into space, their forces undiminished for billions of years.

The median distance of the earth from the sun in its orbit is approximately 150 million kilometers. In comparison to the sun, the earth is very small; its volume is 1.3 million times smaller than that of the sun. Thus only a minute fraction of the radiation from the nuclear reactor sun reaches the earth.

But we owe everything to this radiation.

Without this extraterrestrial energy source there would be no life on earth.

The basic process for the creation and formation of all things living, the transformation of anorganic matter—of carbonic acid and water—into organic substances takes place with the radiation of the light of the sun delivering the energy. This process, called the "assimilation of carbonic acid," supplies the organic building blocks—sugar, carbohydrates, proteins, etc.—for the construction of plants. As no animal organisms can exist without plants, the former need the latter as the food source, the intake of light in the shape of the assimilation process of plants is simultaneously the primary energy source of human life.

Even the formation of the human spirit would have been impossible without the original presence of the light of the sun. The human spirit, our consciousness, represents the highest, most sublime energetic transformation stage of light.

We owe all of the large earthly sources of energy to the extraterrestrial nuclear reactor, the sun:
—the wood of the forests;

—the coal, oil, and gas deposits, in which the sun's warmth of uncounted millions of years was stored;
—the hydraulic power of lakes and rivers—constantly fed by the clouds raised into the sky by the force of the sun—that human intelligence has been able to exploit in a secondary manner in the shape of warmth, light, and electricity.

The extraterrestrial nuclear reactor is also the greatest cleaner and renewer of the life elements i.e. water and air. Based on the action of the sun's heat, pure water rises to the skies from the salt water of the oceans, from polluted rivers and lakes, from the wet ground—and the purified element falls back to earth as refreshing rain or snow that water the plant world.

The sun also supplies the energy necessary for clearning and regenerating the air. During the process of combustion—digesting food in animal organisms, in gasoline engines, in every fire—oxygen is used and carbonic acid is produced. Conversely plants absorb carbonic acid and discharge oxygen into the atmosphere during the assimialtion taking place in the green of the leaf with the light of the sun supplying the energy.

The nuclear reactor sun differs from earthly atomic power plants in that it is:
—absolutely accident and radiation safe;

—poses no threat in disposing of nucelar waste;
—entails neither construction or operating costs;
—has an unlimited supply of fuel, whereas earthly uranium desposits will be depleted in a few decades;
—continuously supplies all peoples of the earth with energy without discrimination;
—has created a green plant world for man and animals that has to give way where earthly atomic power plants are located.

What is man doing when he obtains additional energy through nuclear power plants?

He is kindling sun fires on the earth, that is nuclear reactions, or **physical-chemical** processes of the type taking place on the sun, 150 million kilometers away. The consequences of this incredible distance and a protective earth atmosphere are that only harmless traces of dangerous radiation reach us, while an all-creating, all-sustaining sunlight falls on our plant.

The use of nuclear energy on a large scale (and not to even mention the madness of atomic weapons) creates the danger of contaminating earth with life-threatening radiation. What that means becoms obvious when we consider that life on earth was only possible after nuclear reations here had, with the exception of the traces in the elements that are still radioactive today, died down in the course of billions of years.

The atoms, the construction units of the material world, can be compared to minute solar systems in which electrons circle the atomic core like planets around the sun. With the exception of the processes in the still remaining traces of radioactive elements, all transformations of matter on the planet earth take place in the realm of the electrons, the microcosmic planets, the atomic cores, microcosmically corresponding to the sun, are not damaged.

In contrast, the atomic cores are affected in the case of the nuclear fusion and nuclear fission. In this process, matter vanishes by dissolving into energy. In the planetary reactions-planetary in a macrocosmic and microcosmic sense- in other words during the material transformations in dead matter and during metabolic action in the living organisms of the realms of plants and animals, matter is retained.

Thus the exploitation of atomic energy should not be simply viewed as a further development in the technology of energy production; rather it means something completely new, namely an intrusion into the core of matter, a "de"-velopment away from the natural conditions on which life on our planet is based. From this we can derive that the dangers connected with the exploitation of nuclear energy are life-threatening, and that it is very difficult, if not impossible, to contain them.

Would it thus not have been more reasonable if energy research had concentrated on expanding familiar sources of energy, in other words, on the ones that ultimately originate in the nuclear power plant sun, and that have thus far been able to satisfy our energy needs?

The question whether there will soon be an energy gap that we will have to bridge with nuclear energy is open; it is certain, however, that we will need a new energy concept at some time in the future.

As our present energy supply is largely based on consuming our "capital" of solar energy, or our reserves of oil, gas, and coal, this capital will, as large as it is, be used up in the foreseeable future. Instead of once again propping our future energy supply on a short-lived capital, (namely the uranium deposits that will be depleted in a few decades), we should be striving for an energy plan that restricts itself to the use of "interests", to exploiting the continuously new flow of energy from the nuclear power plant sun. So as to assure that this energy will be sufficient to meet all our respective future needs, we should expand the use of the forces of wind and water, if necessary including other forms of energy that can be collected in the form of "interests". But above all, we should concentrate on an intensified exploitation of the direct radiation of the sun.

It has been calculated that the amount of energy reaching the earth on one single day in the shape of the sun's rays would be sufficient to meet our present energy requirements for several hundred years. Thus the most sensible, worthwhile research projects of our times are the ones dealing with solar radiation as the ideal main source of energy for the future. It is not utopian to assume that human ingenuity will succeed in capturing a small fraction of the vast energy flowing down to us without the aid of wires from our great, safe, inexhaustible, extraterrestrial nuclear power plant, to transform into a useable form, and to thus solve the energy problem for all times to come.

MORE OUTSTANDING BOOKS FROM HUMANICS NEW AGE:

Tao of Management
Bob Messing

An age old study for New Age managers. The Tao of Management will enable managers to see how things happen in their work environment and to understand how energies flow or become blocked. Broad issues of trust, social values, and awareness provide insights into the skills and goals which provide clarity and a sense of accomplishment to the manager.

Tao of Leadership
John Heider

Group leader, program director and teacher, John Heider (Easlen Institute, Human Potential School of Mendocino, Meniger Foundation), explores the Tao Te Ching from the viewpoints of power, potential, and persuasion. From understanding yourself and others, to enhancing creativity and handling conflict, this revealing and important work is for anyone who wants to master the art of effective leadership or who seeks understanding of the nature of things.

Attitudes Make A Difference
Dutch Boling

Boling, a nationally known personality, group leader and business consultant, shares proven techniques for more effective human relations. How to develop better listening and communication skills, creative thinking and problem solving techniques. A new and exciting approach to the power of attitude effectiveness.

These books and other Humanics New Age Publications are available from booksellers or from Humanics New Age P.O. Box 7447, Atlanta Georgia, 30309, 1-800-874-8844. Call or write for your free copy of our publications brochure.